# LIFE IN THE US AIR FORCE

by Douglas Hustad

BrightPoint Press

San Diego, CA

BrightPoint Press


an imprint of ReferencePoint Press, Inc.
Printed in the United States

For more information, contact:
BrightPoint Press
PO Box 27779
San Diego, CA 92198
www.BrightPointPress.com



LIBRARY OF CONGRESS CATALOGING-IN-PUBLICATION DATA

Names: Hustad, Douglas, author.
Title: Life in the US Air Force / by Douglas Hustad.
Description: San Diego : ReferencePoint Press, [2021] | Series: Life in the military | Includes bibliographical references and index. | Audience: Grades 10-12
Identifiers: LCCN 2020003572 (print) | LCCN 2020003573 (eBook) | ISBN 9781682829691 (hardcover) | ISBN 9781682829707 (eBook)
Subjects: LCSH: United States. Air Force. | United States. Air Force--Vocational guidance
Classification: LCC UG633 .H87 2021 (print) | LCC UG633 (eBook) | DDC 358.4/1120973--dc23
LC record available at https://lccn.loc.gov/2020003572
LC eBook record available at https://lccn.loc.gov/2020003573

# AT A GLANCE

- The US Air Force protects the United States in the air.
- The US Air Force is one of the youngest military branches. It was formed in 1947.
- Both male and female members of the air force are called airmen. Some airmen are pilots. Pilots need special training. The process to become a pilot is competitive.
- The process of joining the air force is called enlisting. People can become enlisted airmen or officers. Their position depends on their level of education and training.
- Recruits must go through basic training. This training takes several weeks.

- There are hundreds of different jobs in the air force. Pilots fly various types of planes. Many other airmen support flight operations. They live and work on air force bases.

- In 2019, about 328,000 airmen were active duty. Active-duty airmen serve full-time. Others work in the Air Force Reserve. They serve part-time. The air force offers educational opportunities and other benefits.

- Some airmen are deployed every few years. The air force sends them abroad to serve in other countries.

# UP, UP, AND AWAY

It is a beautiful day for flying. A US Air Force pilot prepares for takeoff. He sits in a Lockheed Martin F-35 Lightning II plane. The F-35 is the most advanced fighter **jet** in the world. It can reach a top speed of 1,200 miles per hour (1,900 km/h).

The pilot lines up on the runway. He pushes the throttle all the way forward.

***The F-35 Lightning II is one of the newest planes in the US Air Force.***

This action quickly speeds up the plane.

The plane reaches takeoff speed in no time.

The pilot guides the plane up into the air. He flies off to complete a top-secret mission.

His goal is to protect the United States.

## WHAT DOES THE AIR FORCE DO?

Many air force **recruits** dream of becoming pilots. But that is just one small part of what the air force does. Even those who do become pilots do not just get to fly all day. It takes a lot of work to keep the air force running smoothly.

Men and women who serve in the air force are called airmen. Some are aircraft maintenance technicians. They make sure each aircraft works well. Mission specialists plan each mission. They help ensure the mission's success. Doctors also work for the air force. They treat airmen and keep

***Some planes, including the C-5 Super Galaxy, are designed to deliver cargo.***

them healthy. Cooks feed airmen. Each of these people has an important role.

The US Air Force is the world's largest air force. It defends the United States in the air.

***A US Air Force pilot flies a C-130J.***

The air force has aircraft in many countries. Airmen keep an eye out for threats to the United States. They serve around the world.

The air force also helps people in times of crisis. Air force pilots fly supplies to people after disasters.

Each air force job requires special skills. Not everybody is cut out for every role. Only a select few can be pilots. But all airmen contribute something. Captain Brian Boardman is an air force pilot. He flies a C-130J transport plane. Transport planes carry supplies or troops. Boardman says, "There are always opportunities to do something different, and experience something you have never done before."[1]

CHAPTER ONE

# HOW DO PEOPLE JOIN THE AIR FORCE?

People join the air force for many reasons. Some people want to be pilots. Others want to get training for a future job. The air force also helps people pay for college. This benefit draws many people to the air force. Others just want to defend their country.

***Recruiters help future airmen find out whether the air force is right for them.***

The process of joining the air force is called enlisting. People who enlist are called recruits. The enlisting process often starts with a visit to a recruiter. Air force recruiters are located around the country. They are airmen. They can answer recruits' questions. They guide recruits through the enlistment process.

People can join the air force as enlisted airmen or officers. This decision depends on their schooling and career goals. Enlisted airmen are ranked lower than officers. Officers lead enlisted airmen. They often have more specialized jobs.

Most people join as enlisted airmen. They do not need a college degree. They must be between seventeen and thirty-nine years old. They must be US citizens or permanent residents. Permanent residents have the legal right to live in the United States.

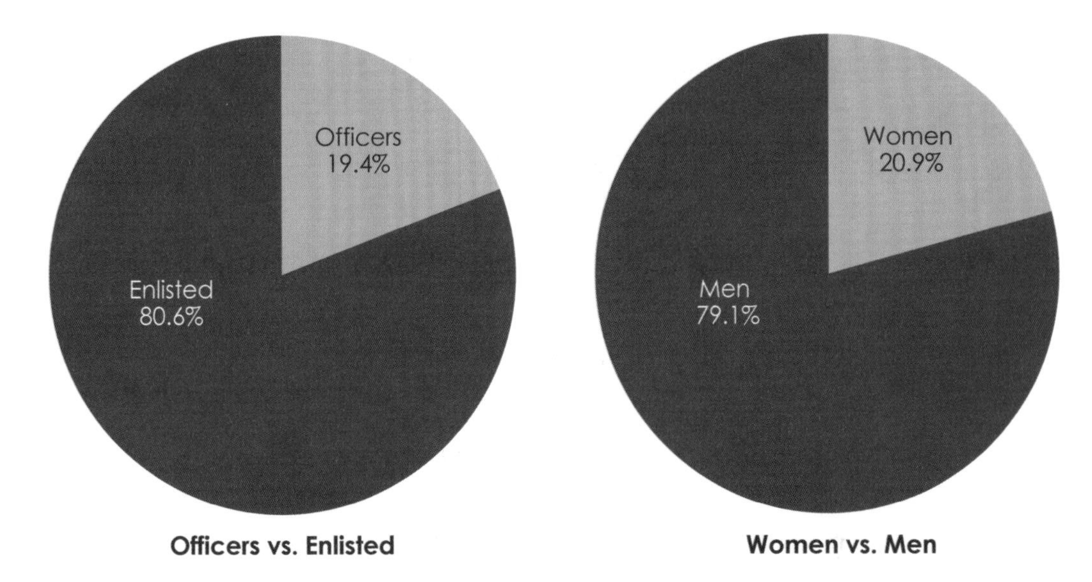


*"Military Demographics," US Air Force, January 1, 2020. www.afpc.af.mil.*

***The US Air Force is made up of more enlisted airmen than officers. More than three-quarters of the people in the US Air Force are men.***

Recruits must also have a high school degree or a General Education Diploma.

People who want to become officers have to meet the same requirements. But they must also have a college degree.

***The US Air Force Academy is known for its famous chapel.***

This can be a degree from any traditional college. Officer recruits do additional training after they earn a degree. Then they can enter the air force.

Officer recruits can go to the US Air Force Academy. This school is in Colorado Springs, Colorado. It is very tough to get into. Students receive training while earning their degree.

## TESTS AND SCREENINGS

All recruits need to take and pass some tests. The tests measure a recruit's abilities. Enlisted recruits must take the Armed Services Vocational Aptitude Battery (ASVAB). The ASVAB is a multiple-choice exam. It tests recruits' knowledge of ten different subjects. The subjects include math, science, and reading. The test also measures recruits' mechanical skills. Some of these subjects are covered in high school classes. Technical subjects may

***The ASVAB is similar to the standardized tests that many high schoolers take.***

require additional knowledge. Recruits are scored in these different areas. Their scores help them find out which jobs might be a good fit.

Recruits must take the ASVAB at a special site. There are many of these sites across the country. Recruits also go through a physical and mental screening. Doctors check recruits' health.

The next step is to talk to a job counselor. Job counselors help recruits find jobs. Recruits identify which jobs they may be good at.

Recruits who pass all the tests and screenings are accepted into the air force. They are assigned a job. Then they move on to Basic Military Training (BMT).

Recruits who want to become officers go through a similar process. They take a special exam. It is called the Air Force

Officer Qualifying Test (AFOQT). The AFOQT is a multiple-choice test. It determines a recruit's interests and abilities.

Next comes a physical and mental screening. This is the same screening enlisted recruits receive. But officer recruits must also go through a review. An air force selection board reviews each recruit. The board meets twice each year. The board members look at each candidate's leadership abilities and education. They decide if a candidate would be a good officer. Recruits who pass these steps are accepted into the air force. Then they go to

Officer Training School (OTS). They receive special training at OTS.

## BASIC MILITARY TRAINING

Enlisted air force recruits go through BMT. This training happens at Lackland Air Force Base (AFB). This base is in San Antonio, Texas. The program lasts about

### LACKLAND AFB

**Training at Lackland AFB is older than the US Air Force itself. The first recruits arrived at Lackland on February 4, 1946. Back then, the air force was still part of the US Army. It was known as the US Army Air Forces. The US Air Force became its own military branch on September 18, 1947.**

eight weeks. Recruits live in barracks. These are big buildings. Recruits sleep in bunk beds. They share bathrooms.

Training starts as soon as recruits arrive at Lackland. They are organized into groups. Each group is called a flight. Instructors lead the groups. The instructors give orders. They stay with the groups throughout BMT.

Recruits go through many physical exercises. The exercises include running and marching. Recruits also receive instruction and technical training. Some training happens in the classroom.

Recruits learn about topics such as first aid. They also train in the field. They practice handling weapons. They learn how to survive in a combat zone. They practice teamwork by working with their flight. Teamwork is important in the air

### ROUTINES IN BMT

**Recruits have routines in BMT. A typical day starts at 4:45 a.m. Recruits exercise for an hour. Then they have breakfast. They also have time to clean their rooms. The rest of the morning is filled with classes and instruction. They have more classroom instruction after lunch. Then, after dinner, recruits have some personal time. They must turn off their bedroom lights at 9:00 p.m.**

***BEAST involves an obstacle known as The Wall.***

force. Airmen must work together to accomplish missions.

The last part of BMT is the most challenging. It is called the BEAST. It happens close to the end of BMT. Recruits learn what life in a combat zone would be like. They live in tents. They must cross a path with hidden devices. The devices are fake explosives. As a group, they must find and safely handle the devices. They also go through obstacle courses. They climb over walls. They crawl up hills on their elbows and knees. They also have to walk on ropes.

Cory Mayo went through BMT in 2011. While it was tough, he enjoyed the training. He said, "BEAST was awesome. The most difficult part of [BMT] is . . . working with your flight to get tasks done on time and correctly. We all have different ways to do things, but it's just a matter of getting everyone to work together."[2]

Recruits who graduate from BMT go through a special ceremony. Their friends and family are invited. They celebrate for a week.

Most enlisted airmen are active duty. They serve full-time in the air force. Part-time service is another option. Recruits who want to serve part-time can join the Air Force Reserve. They also must go through BMT. They must serve at least one weekend per month. They also serve for at least one two-week period each year.

***OTS involves realistic training for missions.***

Reserve airmen may be called upon to serve full-time in an emergency. Then they become active duty.

## OFFICER TRAINING SCHOOL

Officer recruits go through a slightly different kind of training. OTS lasts nine-and-a-half weeks. Recruits must pass physical tests. They do push-ups. They go through obstacle courses. They practice running while wearing gear. They also have specialized training. This training prepares them to become officers.

OTS is divided into four phases. The first phase focuses on teamwork and **discipline**. Recruits study leadership techniques in the second phase. In the third phase, they practice leading groups of people. Recruits receive instruction from air force officers in the final phase. Instructors shout orders at the recruits.

Technical sergeant Paul J. Baisden is an OTS instructor. He says, “One day these officers are going to be tasked to make quick decisions. We . . . prepare them for future real world mission assignments.”[3]

OTS training can be challenging. It prepares recruits for the reality of life in the

**RANKS AND PAY**

**The air force starts paying recruits after they pass their screenings. They get paid the same rate as the lowest-ranking airmen. For enlisted airmen, that rank is E-1. The E-1 rank is also called Airman Basic. E-1 airmen make about $1,700 per month. The lowest rank for officers is O-1. This rank is called second lieutenant. Second lieutenants make about $3,300 per month.**

***A class of OTS graduates attends an award ceremony in 2019.***

air force. Not all recruits make it. Recruits who fail part of the training can go through it again. This process is called recycling.

## BECOMING AIRMEN

Recruits who complete BMT or OTS become airmen. They receive their rank.

Ranks are awarded based on a person's position and experience. Enlisted airmen start out as rank E-1. Officers start out as O-1.

All airmen agree to serve for a certain number of years. Enlisted airmen must serve for a minimum of eight years. At least four of those years must be active duty. Airmen can serve their last four years in the Reserve. Officers must serve a minimum of four years. Pilots have the longest service requirement. They must commit to at least ten years.

CHAPTER TWO

# WHAT JOBS DOES THE AIR FORCE OFFER?

The air force offers more than 200 career options. Some people fly planes, but the branch has many other careers too. There are different jobs available to enlisted airmen and officers. Jobs are divided into several categories. Within each category are many different jobs.

***In technical training airmen might learn how aircraft work and how to fix them.***

After graduating BMT, enlisted airmen enter technical training. The length of this training period depends on the type of job. It can last anywhere from six to seventy-two weeks. Airmen go to class for eight hours each day. They are in the classroom five days each week. They get hands-on training. This training helps them become an expert in their job field.

Many air force jobs are specialized. Some airmen learn how to work on aircraft. There are also many support jobs. People in support roles assist airmen. For example, some people are doctors or lawyers. They work for the air force.

## AIR FORCE PILOTS

Many air force recruits set out to become pilots. The process to become a pilot is very competitive. Pilots must meet the requirements before age thirty-three.

They must have no history of asthma or allergies as an adult. And they must have good vision with no eye problems. Pilots can be any height. But pilots who are tall may not fit in all types of aircraft.

Only officers can become pilots. They must take the AFOQT. Part of this test assesses a person's flying skills.

**AIR FORCE ONE**

**The US Air Force maintains the US president's personal plane. The plane is commonly known as Air Force One. It is only officially called that when the president is on board. The plane can fit up to seventy passengers. Air force pilots fly the plane. It is based out of Andrews Air Force Base in Maryland.**

***A pilot gets ready to climb into his F-22 Raptor fighter jet.***

Another part assesses navigation skills. Candidates who pass can be admitted into Air Education and Training Command (AETC). This program is in San Antonio,

Texas. Some airmen do not have their pilot's license. They can train to get this license in AETC. Then they receive specialized flight training.

A pilot's main job is to carry out missions. Some airmen are fighter pilots. They fly in combat missions. Some air force planes can carry troops. Others transport equipment. Each pilot specializes in flying a certain type of aircraft.

Pilots do more than just fly. Some give instructions to other pilots from the ground. Others help organize and prepare for missions.

## WORKING WITH AIRCRAFT

Some aircraft do not have pilots on board. Instead, pilots control the aircraft from the ground. In the air force, this type of aircraft is called a Remotely Piloted Aircraft (RPA). Some people also call them drones. RPAs can fly missions without putting pilots in harm's way. They can deliver supplies and weapons. They can also take photos and videos. Airmen may use RPAs to spy on enemies.

RPA pilots must be officers. They go through training at Randolph Air Force Base. This base is in Texas. The training takes nine-and-a-half weeks. RPA pilots must know how to navigate and fly aircraft. They learn about different types of missions. RPAs can be piloted from anywhere. More and more missions are being designed for RPAs. For example, an RPA can be used with another aircraft. It can

**PRESIDENTIAL AIRMEN**

**Two US presidents have served in the air force. Ronald Reagan served in World War II (1939–1945). At that time, the air force was called the US Army Air Forces. Reagan was an actor. He produced training films for the military. President George W. Bush also served in the air force. He was a pilot in the Texas Air National Guard in the 1970s. The Air National Guard is part of the Air Force Reserve.**

fly ahead of the aircraft. It makes sure a location is safe before any airmen get there. RPAs can also fire weapons.

Enlisted airmen cannot become pilots. But they may still have jobs on board aircraft. Some airmen are pararescue jumpers (PJs). They jump from aircraft. They use parachutes. They can access hard-to-reach places. PJs can treat injured airmen. They also help rescue people. Master Sergeant Ivan Ruiz is a PJ. He says, "[PJs are] very experienced with aircraft. . . . On top of all that we're divers and shooters, and we have medical capability. We're kind

***Pararescue jumpers train using their MV-22 Osprey aircraft.***

of the out-of-the-box thinkers of the military team."[4]

Airmen may need special supplies while on a mission. For example, they may fly to high **altitude**. The air is thin at these levels. Airmen may need to use oxygen masks. These masks deliver oxygen to help airmen breathe. Some airmen are in charge of taking care of this equipment. They have an important job. Without working supplies, missions could fail.

Other airmen help maintain and fix aircraft. Aircraft have many parts. Many airmen are needed to keep aircraft in working order. Planes also require a lot of fuel. Some airmen have the important job of refueling planes on the ground and in the air. Aircraft may also carry bombs or other

***Loading weapons onto aircraft is an important, delicate job.***

weapons. Weapons experts load bombs onto planes.

## OTHER JOBS

Part of the air force's mission involves **cyberspace**. Cybersecurity experts help keep military computers safe from threats.

Some work as cyberspace operations officers. They are technology experts. They use technology to plan missions. They also make sure the air force is not vulnerable to a cyberattack. They must keep up with the latest technology. They need to stay one step ahead of the nation's enemies.

The US Air Force is a huge organization. It takes hundreds of thousands of employees to keep it running. Doctors and dentists keep airmen healthy. Cooks make sure airmen get healthful meals. Some people are in the US Air Force Band. They play music at air force ceremonies.

***An air force photographer prepares to take photos from inside an airplane.***

Air force journalists take photos and videos. They share news and information with the public. All of these people have important roles in the air force.

CHAPTER THREE

# WHAT IS LIFE LIKE IN THE AIR FORCE?

The air force leads missions all around the world. It has bases in thirty-five US states. There is also an air force base in Washington, DC. The air force has eight bases in other countries too. These bases are spread across friendly countries in Europe and Asia.

***Tinker Air Force Base is in Oklahoma.***

Airmen have some choice in where they are assigned. They make a list of their sixteen preferred bases during training. They get their assignment near the end of their technical training. The air force takes their interests into account. But where an airman gets stationed depends on the needs of the air force. The air force may have a need for a particular job in one part of the world. If an airman has this job, that is probably where she is going to go.

## LIFE ON BASE

Most airmen live and work on base. Bases are like small cities. People get up and go to work on the base each day. They typically work eight hours per day. Sometimes they may need to work longer. Their workload depends on the air force's needs.

Single and low-ranking airmen live in barracks. On-base housing is provided free of charge. Airmen can move off the base once they reach a certain rank. They may live in apartments near the base. The air force helps pay for their living expenses.

Some bases have houses available for airmen with families. These areas are like neighborhoods. Most bases have schools

**MUSTACHE MARCH**

**An air force base can be a hairy place come March. This month is known as Mustache March. Many male pilots grow out a mustache. They do it to honor one of the greatest pilots in air force history. Brigadier General Robin Olds was a fighter pilot. He served for thirty years. He fought during World War II and the Vietnam War (1954–1975). He shot down seventeen enemy planes. He was well known for his bushy mustache.**

for the airmen's children. Some have playgrounds and youth sports teams to join.

There are plenty of activities available for airmen and their families. Bases have workout facilities. Some bases have golf courses for members of the military. There are also pools and recreation centers. Airmen earn thirty days of paid vacation each year.

Medical services are free of charge. Hospitals and pharmacies are available on base. Pharmacies give people medication. Dentists are also available on base or at a nearby office.

## EDUCATIONAL OPPORTUNITIES

Many people are drawn to the military because it helps them get an education. Military branches often offer some type of education. The education can include technical training. The military also helps **veterans** pay for college.

### THE ROOF STOMP

**A roof stomp is a type of air force celebration. It can happen whenever there is something to celebrate. For example, airmen may do a roof stomp to welcome a new commander. Airmen go over to the commander's house. They jump on the roof. They make noise until the person comes outside. Then they all go in for a party.**

The air force offers educational opportunities. Airmen can go through a program called the Community College of the Air Force (CCAF). This program is offered at 112 air force schools around the world. Airmen can learn more about their career field. They can earn a degree. The CCAF offers two-year degree programs. It awards 22,000 degrees each year. The CCAF is the world's largest community college system.

Some high school students know they want to join the air force. But they want to go to college first. There are ways to

***A group of airmen celebrate earning their CCAF degrees in December 2019.***

get involved in the air force while going to school. College students can go through a special program. It is called the Reserve Officers' Training Corps (ROTC). ROTC students can earn scholarships to help pay for school. They get specialized training to prepare them for the air force.

ROTC students attend traditional colleges. But they study and train alongside other ROTC students. They develop leadership and management skills. They commit to serving in the air force when they graduate. They enter the air force as officers.

The US Air Force Academy also prepares college students for the air force. This school is very hard to get into. Unlike most colleges, students do not apply for admission. They must be nominated.

***US Air Force Academy cadets visit air force bases to learn from active-duty airmen.***

Most applicants write to their representative in the US Congress. Applicants can also seek a nomination from the vice president.

Angela Maria Villarreal went to the US Air Force Academy in the late 2010s. The workload was challenging. But she enjoyed the training. She said,

> *The best thing about this place is the opportunities that come with it. I've done so many things since I've been here. I've jumped out of planes. I've worked with live birds of prey. My choice of coming to the Air Force Academy will definitely affect the rest of my life.*[5]

The US Air Force Academy is free. Students become officers after they

***An airman works on an online college course during a long flight.***

graduate. They must then serve in the air force. The required length of service depends on the student's career.

## SPECIALIZED EDUCATION

Not all airmen have a college degree. Some want to earn a degree while serving.

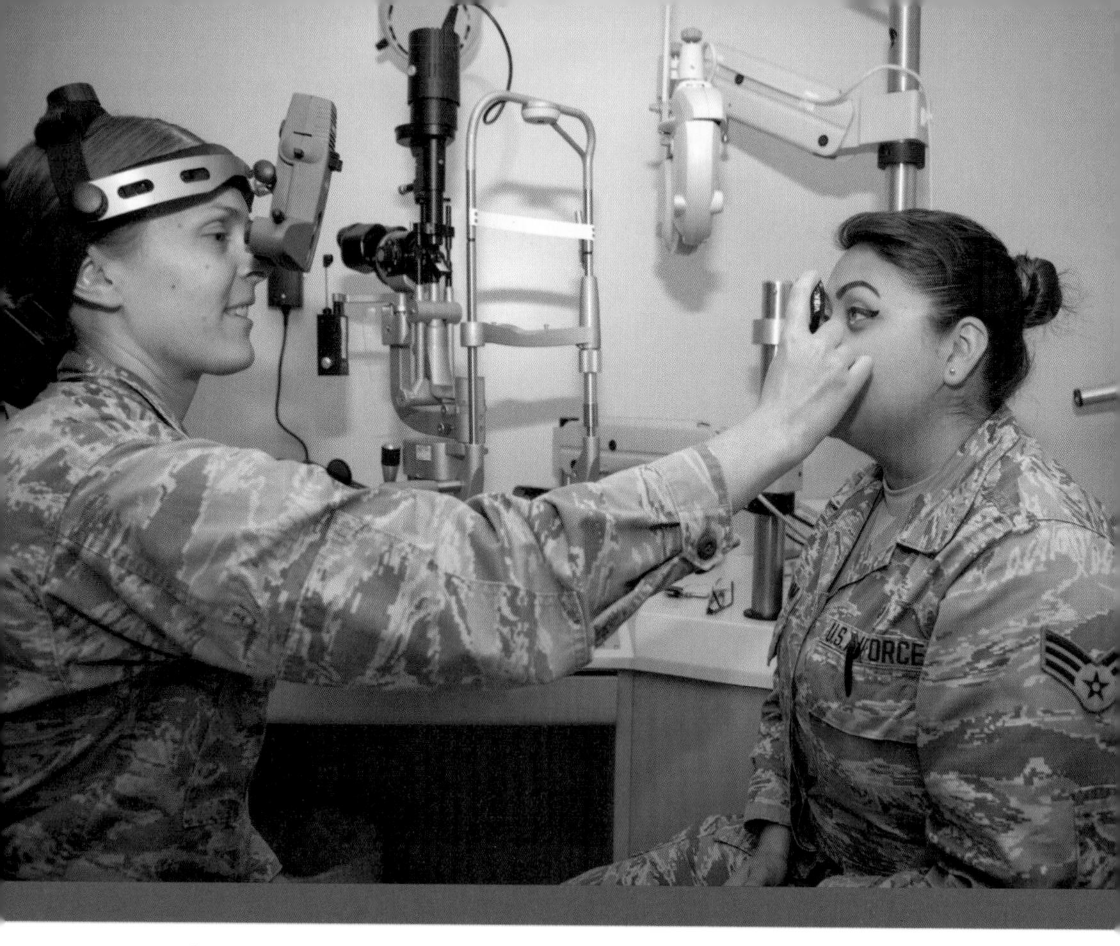


***Airmen can have their medical education paid for by the air force.***

Others choose to go to college after their service ends. The air force can help these people get an education. The Post-9/11 GI Bill was signed into law in 2009. It applies to people who have served in the military since September 10, 2001. These people must have served for at least ninety days in a row. The bill pays for public college. It also helps pay for living expenses, books, and supplies.

The air force offers many programs for specialized education. Captain Ryan McHugh is a doctor. He works for the air force. He treats airmen's children. He says, "I went to medical school at a **civilian** school, and the air force paid for it. You also have the option of going to the military medical school in Bethesda, Maryland. So, there's kind of two different ways."[6]

After medical school, future doctors must complete a residency. A residency is a training period at a hospital. The air force may provide airmen with money during a residency.

The air force also has a program for law school. Students can get up to $65,000 in loans to help pay for law school. Additionally, there are programs for airmen who want to become religious advisors, or chaplains. These are skills people can put to use long after they leave the air force.

There are lifelong benefits for those who make the air force a career. Some airmen

***In 2019, Second Lieutenant Saleha Jabeen became the first female Muslim chaplain in the US Air Force.***

serve for twenty years or more. Then they retire, or leave the air force. They receive payments from a retirement plan. Veterans may also qualify for other benefits. These benefits include health care and loans to help them buy a house.

CHAPTER FOUR

# WHAT IS DEPLOYMENT LIKE?

Sometimes the air force sends airmen around the world. This can happen in a time of war. Or the air force may have a mission in another country. The mission may require a lot of airmen. The process of sending airmen abroad is called deployment. Deployment is temporary.

***An airman prepares to board a C-130 for deployment to Southwest Asia.***

Airmen serve abroad for a certain period of time. Then they return to their home base.

Not every airman will deploy. It depends on the needs of the air force. It also depends on the airman's job.

But deployment is a possibility for all airmen. Most airmen are deployed at least once.

Airmen have different duties during deployment. Some may be involved in combat. Others may have support roles. Captain Maria DeArman is an airman.

**AIR FORCE ASTRONAUTS**

**Many astronauts have served in the air force. Michael Collins was an air force general. He flew the Apollo 11 spacecraft around the moon. Buzz Aldrin was also a member of the Apollo 11 crew. He was the second person to walk on the moon. He was an air force fighter pilot. Guion Bluford was the first African-American in space. He served as an air force colonel.**

She works as a doctor. She was deployed to Afghanistan. Her deployment lasted one year. Her role was to help people in Afghanistan. DeArman said, “We’re setting up roads and helping to build hospitals. . . . I got to teach midwifery, how to deliver babies. I got to [work] with Afghan doctors.”[7]

Some airmen are deployed to help with disaster relief. They help people recover after a natural disaster such as a hurricane. In 2019, a cyclone hit Beira, Mozambique. The cyclone’s winds whipped at 105 miles per hour (169 km/h). The storm destroyed

***Airmen load cargo for a disaster relief mission to Puerto Rico after an earthquake struck the island.***

people's homes. It flooded the city. Nearly 2 million people needed help after the storm. The US Air Force sent teams to help. The airmen brought food, medicine, and other supplies.

## CHALLENGES

Deployment can be exciting. Airmen get to put their skills to use. Many people join the air force because it allows them to travel and live around the world. Travel is a way to have new experiences. But it also comes with challenges. Airmen must learn new customs. Customs are the common behaviors and attitudes of people living in a certain place.

The length of deployment varies. It usually lasts between four and twelve months. The air force tries to give airmen several months of notice. It also tries to take

***Mail service at Ramstein Air Base in Germany keeps airmen connected with family and friends at home.***

an airman's family situation into account. But nothing is guaranteed. In times of war, airmen could be deployed with very little notice.

Airmen can keep in touch with their families while abroad. There are chances for airmen to call or write home. But the mission comes first. Airmen may not be able to communicate for long periods of time. They may also be deployed to a

### THE US SPACE FORCE

The air force was the newest US military branch until December 2019. Then the US Space Force (USSF) was created. The USSF carries out missions in space. Before the USSF was founded, the Air Force Space Command was in charge of space-related missions. It used technology such as **satellites** to gather information. It sometimes helped plan or carry out attacks. All airmen from that command were reassigned. They now work for the USSF.

secret location. Then they cannot tell their families where they are deployed. But families can still send them letters. People can mail letters to an air force office. The office delivers the letter to the airman.

Video chat is a common form of communication. It allows airmen to see and chat with their families. One air force spouse says, "[Video chat] has probably saved marriages and families. It takes away so much separation."[8]

## TRAVEL AND TIME OFF

An airman's mission may change often. Sometimes an airman is needed elsewhere.

***Video chat is an easy way for airmen to keep in touch with families back home.***

She may have to move to a new base. An airman may be relocated. She may be sent to another US base. Or she may be sent to a base overseas.

***Airmen around the world work together to make sure the US Air Force can continue to fly and fight.***

Airmen may also travel for fun. The air force gives them some time off. Airmen get thirty paid vacation days each year. Some airmen spend that time traveling with loved ones. It is easy for airmen to travel. They can fill any empty seats on air force flights. They have access to many flights and locations all around the world. Many airlines give military discounts too. They may also offer discounted flights or baggage fees to an airman's family.

Serving in the air force is hard work. But the air force offers many opportunities. All airmen contribute to the air force's success. They play a key role in defending the United States.

# GLOSSARY

**altitude**

height above the ground

**civilian**

someone who is not in the military

**cyberspace**

having to do with the internet

**discipline**

carefully controlled behavior

**jet**

a type of aircraft engine in which burning fuel makes a stream of gas that propels the plane forward

**recruits**

people who are new to the military

**satellites**

machines that fly in space around Earth and have a useful purpose

**veteran**

someone who served in the military

# SOURCE NOTES

## INTRODUCTION: UP, UP, AND AWAY

1. Quoted in "A Day in the Life of a Pilot," *Little Rock Air Force Base*, September 4, 2014. www.littlerock.af.mil.

## CHAPTER ONE: HOW DO PEOPLE JOIN THE AIR FORCE?

2. Quoted in Amy McCullough, "BMT Gets Real," *Air Force Magazine*, January 18, 2020. www.airforcemag.com.

3. Quoted in William J. Blankenship, "The Air Force's Future: A Day in the Life of an OTS Trainee," *Maxwell Air Force Base*, July 19, 2012. www.maxwell.af.mil.

## CHAPTER TWO: WHAT JOBS DOES THE AIR FORCE OFFER?

4. Quoted in Mark Synnott, "Injured Behind Enemy Lines, This Guy Is Your Best Friend," *National Geographic*, August 18, 2016. www.nationalgeographic.com.

## CHAPTER THREE: WHAT IS LIFE LIKE IN THE AIR FORCE?

5. Quoted in "Day in the Life," *Air Force Academy Admissions*, February 24, 2015. www.youtube.com/watch?v=np1dxwbrb8c.

6. Quoted in "Paying for College," *Today's Military*, n.d. www.todaysmilitary.com.

## CHAPTER FOUR: WHAT IS DEPLOYMENT LIKE?

7. Quoted in Maria DeArman, "Family Medicine Physician," *Today's Military*, n.d. www.todaysmilitary.com.

8. Quoted in "Communicating with Your Partner on Deployment," *Military.com*, n.d. www.military.com.

# FOR FURTHER RESEARCH

## BOOKS

Roberta Baxter, *Work in the Military*. San Diego, CA: ReferencePoint Press, 2020.

Sam Caulkins, *My Uncle Is in the Air Force*. New York: PowerKids Press, 2016.

Lee Slater, *Pararescue Jumpers*. Minneapolis, MN: Abdo Publishing, 2016.

Brandon Terrell, *Guarding Air Force One*. Mankato, MN: The Child's World, 2016.

## INTERNET SOURCES

"The History and Roles of the Air Force," *Military.com*, n.d. www.military.com.

"Military Careers," *US Bureau of Labor Statistics*, September 19, 2019. www.bls.gov/ooh/military.

"US Air Force Academy by the Numbers," *US Air Force Academy*, n.d. www.usafa.edu.

## WEBSITES

### National Museum of the US Air Force (USAF)

**www.nationalmuseum.af.mil**

The National Museum of the USAF is the official air force museum. It is on Wright-Patterson Air Force Base near Dayton, Ohio. The museum has exhibits that explore the air force's history and how it works today.

### US Air Force

**www.airforce.com**

The official US Air Force website shares information about joining and serving in the air force. Visitors can learn more about air force careers.

### US Space Force

**www.spaceforce.mil**

The US Space Force is the newest part of the US military. It is organized under the US Air Force. Its website offers information about the new branch's mission.

# INDEX

# IMAGE CREDITS

Cover: Tech. Sgt. Kat Justen/US Air Force/DVIDS
5: Airman 1st Class Ryan C. Grossklag/US Air Force/DVIDS
7: Capt. Kip Sumner/US Air Force/DVIDS
9: Airman 1st Class Ryan C. Grossklag/US Air Force/DVIDS
10: Senior Airman Elizabeth Baker/US Air Force/DVIDS
13: Airman 1st Class Ericka A. Woolever/US Air Force/DVIDS
15: © Red Line Editorial
16: Mike Kaplan/US Air Force/DVIDS
18: © panitanphoto/Shutterstock Images
24: Tech. Sgt. Katherine Spessa/US Air Force/DVIDS
27: Airman 1st Class Charles Welty/US Air Force/DVIDS
30: Trey Ward/US Air Force/DVIDS
33: Airman 1st Class Madeleine E. Remillard/US Air Force/DVIDS
36: 2nd Lt. Sam Eckholm/US Air Force/DVIDS
41: 1st Lt. Christin St. John/US Air Force/DVIDS
43: Senior Airman Rhett Isbell/US Air Force/DVIDS
45: Crosby Shaterian/US Air Force/DVIDS
47: Tech. Sgt. Lauren Gleason/US Air Force/DVIDS
53: Staff Sgt. Jordan Hohenstein/US Air Force/DVIDS
55: 2nd Lt. Danny Rangel/US Air Force/DVIDS
57: Tech. Sgt. Dan Heaton/US Air National Guard/DVIDS
58: Airman 1st Class Scott Warner/US Air Force/DVIDS
61: Tech. Sgt. Armando A. Schwier-Morales/US Air Force/DVIDS
63: Staff Sgt. Jon Alderman/US Air Force/DVIDS
66: Tech. Sgt. Joe Harwood/Ohio Air National Guard/DVIDS
68: Airman 1st Class John R. Wright/US Air Force/DVIDS
71: Tech. Sgt. Louis Vega Jr./US Air Force/DVIDS
72: Airman 1st Class Matthew Seefeldt/US Air Force/DVIDS

# ABOUT THE AUTHOR

Douglas Hustad is the author of several books on science and technology for young people. Originally from the San Francisco Bay Area, he now lives in North County San Diego with his wife and their two dogs.